GOODNIGHT WORLD

A LETTER FOR AIR

To my sons Creedence and Cruz,

and to children everywhere.

May you always shine bright and follow your dreams.

Goodnight World, A Letter For Air
is the third book of the series,
Goodnight World.

In this series, you will follow a young girl as she writes in her journal every night before bed sharing her thoughts on the importance of all living things on this Earth. She takes you on a journey that will ignite your imagination and deepen your connection to the world around you.

A Letter For Air shows an appreciation for being in nature and breathing in fresh air. It encourages you to look deeper into what causes air pollution and how you can make a difference in your own life. Let yourself be taken away through whimsical illustrations and poetic verses that unlock your curiosity and passion for a cleaner Earth.

Visit our website at
laceyboucherbooks.com

or check out our Instagram
@laceyboucherbooks

Lacey Boucher Books is more than just books, it's a vibe!

We look forward to connecting!

This book belongs to:

Contact the publisher at

lacey@laceyboucherbooks.com

Paperback ISBN 978-1-7781-859-46

Ebook ISBN 978-1-7781-859-53

Stay connected with us at

laceyboucherbooks.com

GOODNIGHT WORLD

A LETTER FOR AIR

Written by Lacey Boucher

Illustrated by Eilnaz Barmayeh

GOODNIGHT WORLD

A LETTER FOR AIR

The stars fill up the sky as we camp by the lake.
The peacefulness and beauty keep my eyes wide awake.

I take in the view and breathe the fresh air.
From the city to nature, the difference is there.

The sky is so clear in the day and the night.
The air smells of woods and the stars shine so bright!
I have always loved being away from it all.
The cars, the factories, and the buildings so tall.

Sometimes in the city there is something called smog.
A type of air pollution that fills the sky like a fog.
Smog is caused by people burning oil and coal.
Carbon is produced and our ozone pays a toll.

So much of this carbon is put in our atmosphere.
But we notice firsthand when the sky is not clear.

We can all do more and reduce our carbon print.
By leaving our cars and taking a sprint!
A sprint to the finish line of finding a solution.
Using clean energy that helps stop air pollution.

So goodnight World, I will dream of what I can do.
Turn off lights, ride a bike, I will start with those two!

GOODNIGHT WORLD

A LETTER FOR AIR

laceyboucherbooks.com